SOTHIS PRESS

Sothis Press aims to completely reimagine the public domain book with an emphasis on universal access to captivatingly designed print editions. We believe that technology will enable every single book ever published to be perpetually accessible in beautiful print form. Sothis is beauty in ubiquity.

www.sothis.press
1 Waterhouse Square, London, EC1N 2ST

A Tree Hurts, Too

Alex L. Shigo

SOTHIS PRESS

ISBN 978-1-0365-0927-9

This edition is published by
Sothis Press © 2023. The company
is registered in England and Wales.
The Company Number is 14900389.

A tree hurts, too

Agriculture Information Bulletin No. 396

U.S. DEPARTMENT OF AGRICULTURE FOREST SERVICE

This booklet outlines what happens most of the time as decay develops in a living tree. The drawings are designed to give an accurate general presentation of the decay process by focusing only on the major portions of a extremely complex process that involves the interactions among microorganisms, environmental factors, and th tree. he better we understand these processes the better we can prevent and control decay—the major cause of damage to all species of trees throughout the world.

The information and illustrations are based on the Pioneer Forestry Research Project of Dr. Alex L. Shigo, a Forest Service Plant Pathologist, stationed at the Forestry Sciences Laboratory in Durham, New Hampshire. This booklet was originally issued by the Northeastern Forest Experiment Station, Forest Service, U.S. Department of Agriculture, Upper Darby, Pa., in 1973 under the same tile.

"A Tree Hurts, Too" is part of a Forest Service information program designed to make forestry research results more understandable and usable for practitioners and laymen.

The illustration on the cover illustrates how the fruit body of a fungus on a large old basal wound indicates internal decay as shown by the cut-away section.

Art work is by David M. Carroll, Illustrator

A tree, like a human person, can be hurt . . . Hurt in the sense of injury, damage, or wounds, but of course not in the sense of feeling pain. And like a human person, a tree has ways of healing its wounds and curing—or at least living with—the wound diseases that afflict it.

A tree is unique among all living organisms. It lives longer and it grows taller and larger in mass than any other organism in the sea or on the land. Large as it is, a tree can be wounded rather easily.

And like you and me, a tree responds to its wounds. It may not say "ouch," but it does react. A tree "hurts," too.

The trouble usually begins with a wound, a wound that penetrates the bark and exposes the wood beneath. A tree wound can be caused by a bird, an animal, an insect; by fire, by storm damage; by a logging machine or an automobile bumping against it; or by a nail being driven into it.

The tree reacts. A chemical reaction takes place in the tree to protect the wound. If the tree is healthy—it heals its wound, and no damage is done.

But often the wound provides an entry for the organisms that cause disease, bacteria, fungi, and viruses. Once these organisms get into the tree, a very complex succession of events takes place.

The tree may die; it may decay inside; or it may suffer no more damage than a discoloration of its wood.

This booklet was prepared to show how decay develops in living trees, to show the marvelous ways in which a tree reacts against the agents of wound diseases, and to tell you what you can do to prevent damage to trees.

TREES are well adapted for life on earth. In some ways, they are superior organisms. Trees live longer and grow taller and larger in mass than any other living thing on the land and in the sea.

Compare the size and longevity of
MAN
to
Largest land animal—Elephant—12 feet tall
Largest sea animal—Blue whale—100 feet long; 150 tons
Tallest sea plant—Giant kelp—300 feet long
Oldest tree—Bristlecone pine—5,000 to 6,000 years old.

Tallest and largest trees
Redwood and Sequoia—over 325 feet tall; 2,000 to 3,000 years old.
(All drawings are to scale.)

Redwood
Sequoia
Bristlecone Pine
Kelp

*But, despite their fine adaption to life on this planet, they too—
like all organisms—*

Grow;

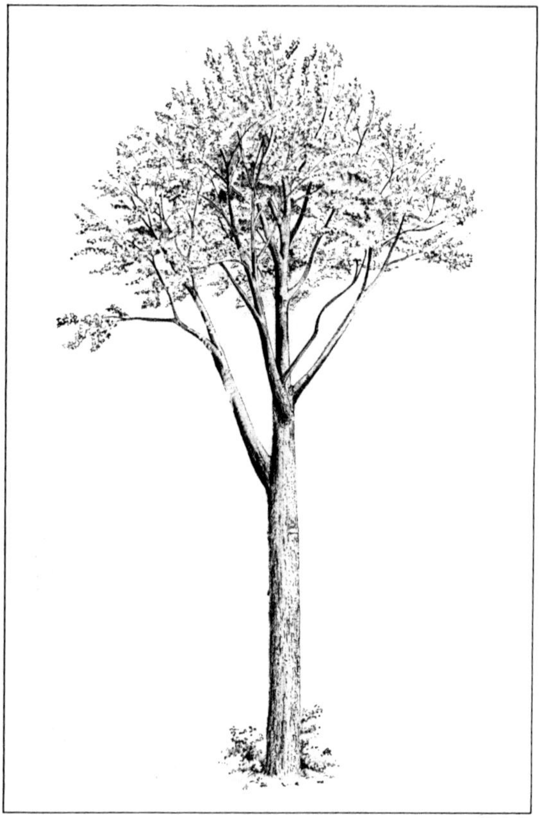

Wane;

Die;

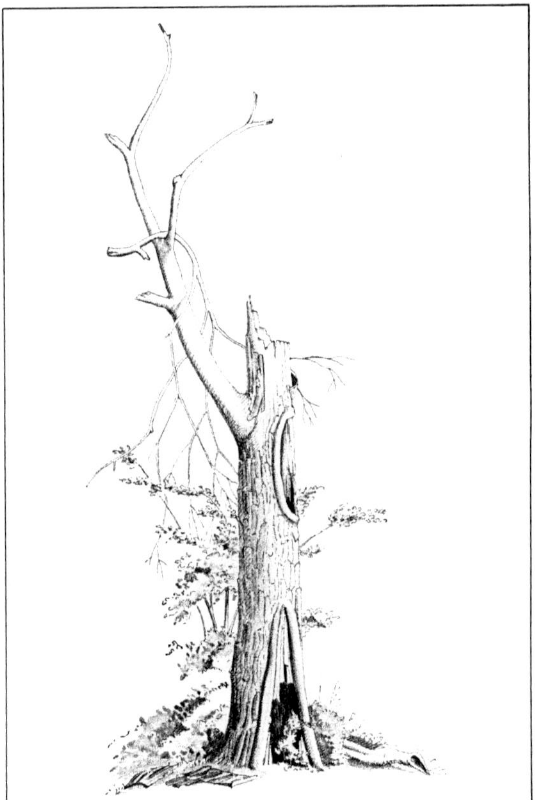

Decay.

TREES die from many causes—diseases, pollution, fire, environmental extremes, insects, animals, etc.

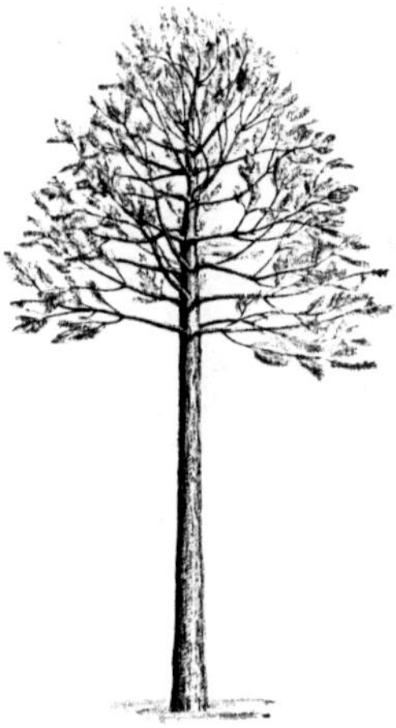

Sick Pine

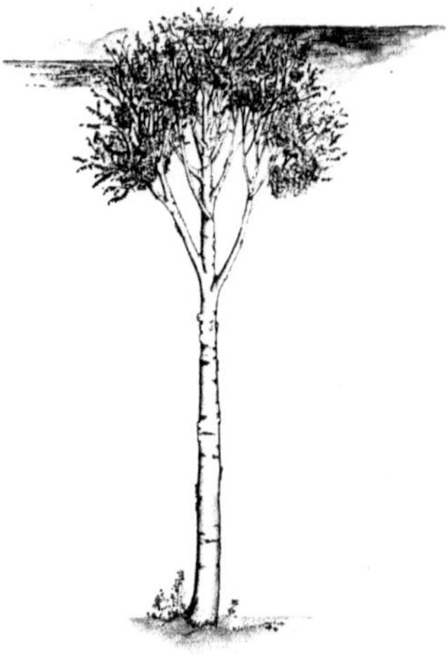

Sick Birch

Yet a major cause of DEFECT, DEATH, and DECAY of all species of trees throughout the world is infection of WOUNDS by microorganisms.

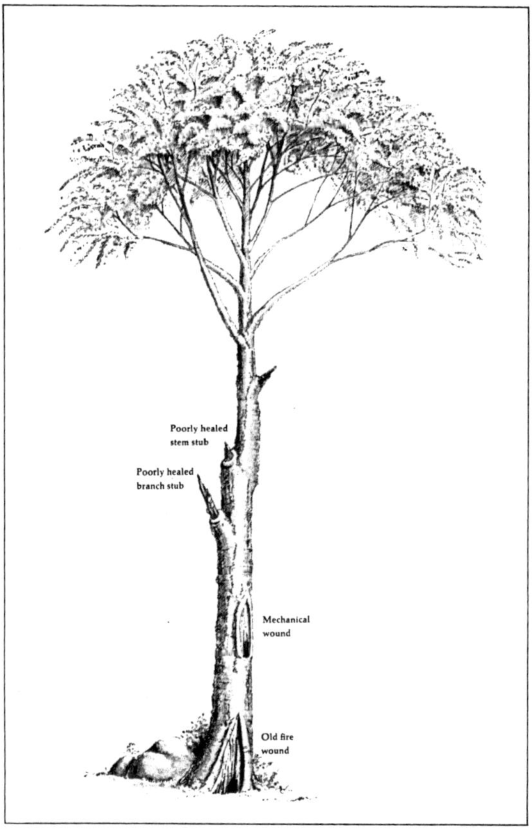

Some microorganisms are beneficial; some are destructive. Their life-span is measured in days, and their size in thousandths of an inch. But without them, no organic matter would decompose, there would be no reuse of once-living materials, and thus there would be no continuation of life.

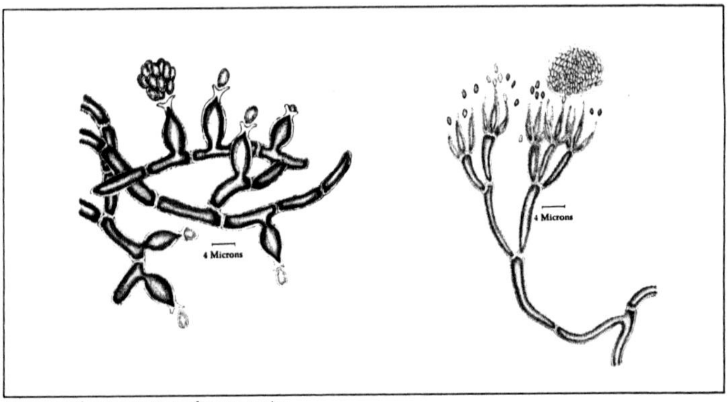

Microscopic fruiting structures of two aggressive pioneer fungi that invade wood through wounds.

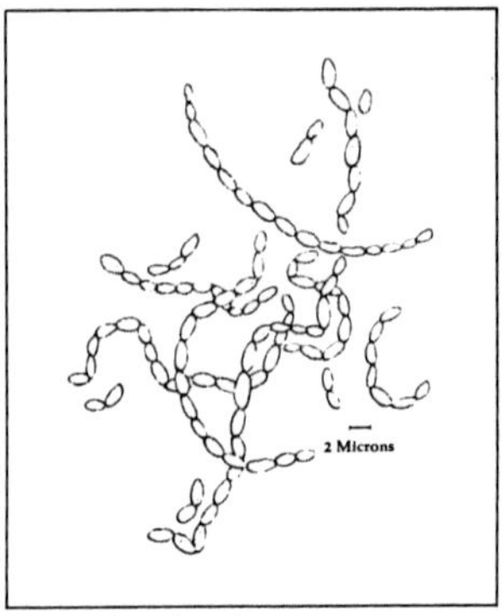

Wood-inhabiting bacteria.

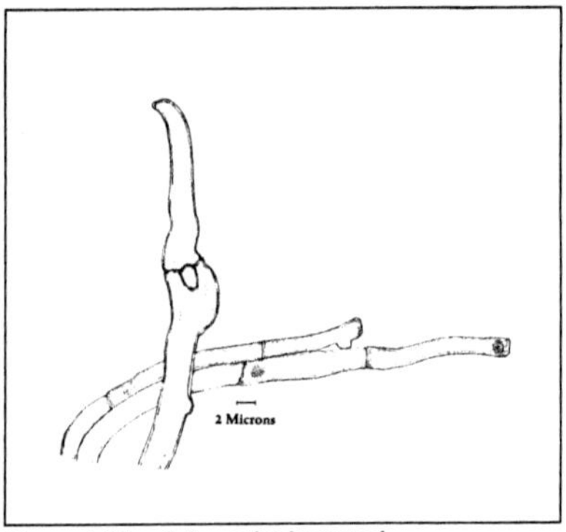

Microscopic tubes of a fungus that decays wood.

Mushroom fruit bodies of a fungus that decays wood.

Perennial shelf-like fruit body of a wood-decay fungus.

Trees live longer and grow taller and larger than other organisms mainly because they have very effective ways of resisting death and decay. Over millions of years, trees have developed effective built-in systems of protection against wounds and infection by microorganisms.

Vigorous trees such as these have strong systems for protecting themselves and repairing their wounds.

Microorganisms that cause decay sometimes surmount the protective barrier set up by the tree after wounding. When the tree's vigor is high and the wounds are minor, the advantage tips toward the tree. When the tree's vigor is low and the wounds are severe, the advantage tips toward the invading microorganisms. The cumulative effects of many minor stress factors over time also tip the balance in favor of the microorganisms.

Vigorous tree—balance in favor of the tree. Weak tree—balance in favor of microorganisms.

A TREE should be thought of as a multiple plant, rather than as a single one, for the cambium—the layer of tissue one cell thick between the wood and the bark—produces a completely new layer of wood and bark tissue every growing season. In a sense, a new tree envelops the old one every year.

Anatomy of a red oak.

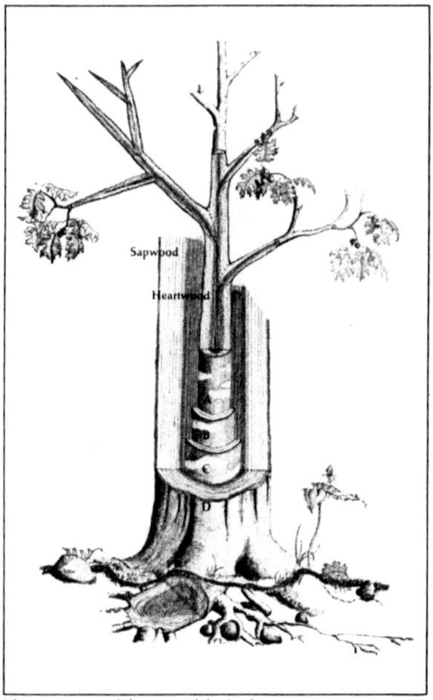

The appearance of the tree with bark when it was
A—8 years old. B—10 years old.
C—12 years old. D—22 years old.

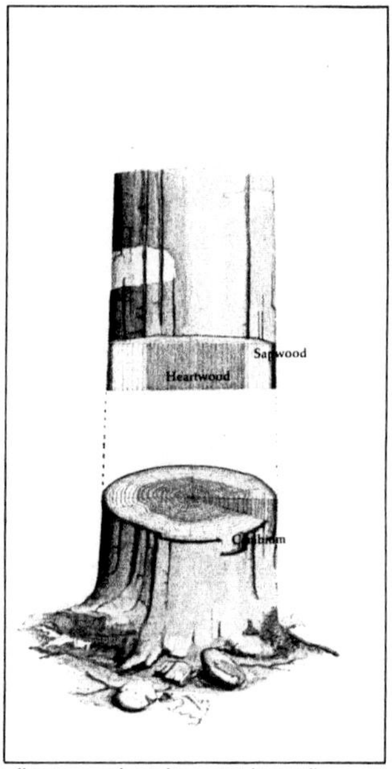

yellow—sapwood; wood containing living cells.
brown—heartwood; wood altered by natural aging processes; no living cells.
green—cambium; living cells that produce new tissues.

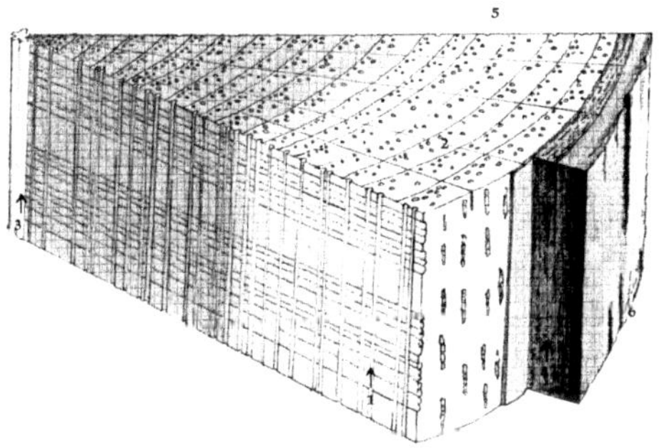

1—vessels.
2—ray cells—living cells in sapwood.
3—pith—center of tree.
4—inner bark or phloem; materials flow downward mostly.
5—wood or xylem; in sapwood materials flow upward mostly.
6—outer bark—protective shield.

The chemical barriers formed in the wood around the wound inhibit invasion by most microorganisms, but certain ones sometimes can grow through the barrier. When this happens, a second wave of different microorganisms moves in behind the first invaders. Still others follow. A SUCCESSION of microorganisms is involved in the processes that result in decay.

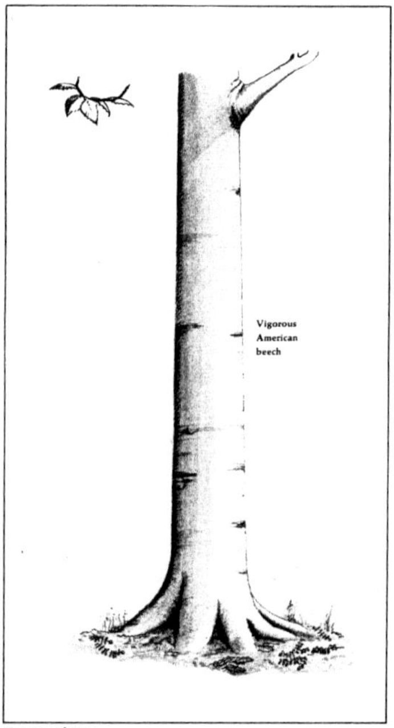

Uninjured tree

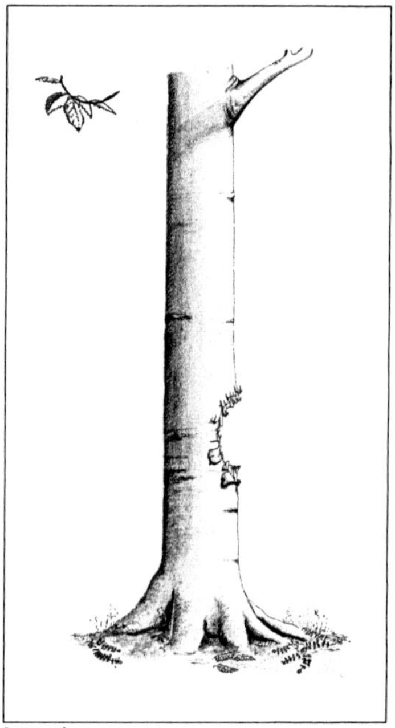

Injury at base

Color key to all drawings

RED—Host response to wounding, chemical reactions—to STOP invasion.

GREEN—Invasion processes by microorganisms—to GO through the barriers set up by the tree.

BLUE—Infection of dead and dying tissues on wound surface.

BROWN—Decay processes.

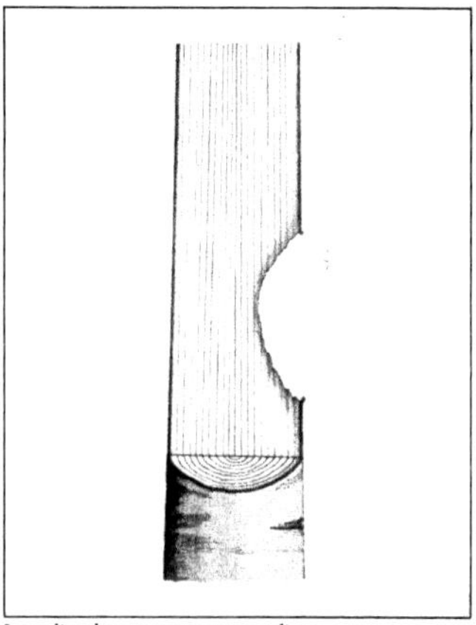

Immediate host response to wounding.

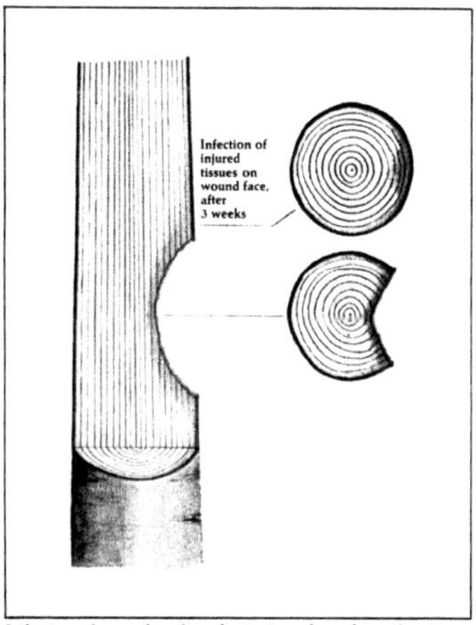

Infection of wound surface; bacteria and nondecay fungi.

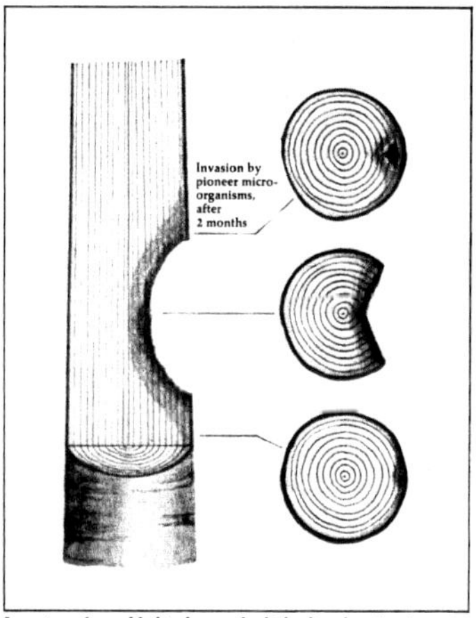

Invasion of wood behind wound; slight discoloration in GREEN area. Invasion by pioneer bacteria and nondecay fungi usually.

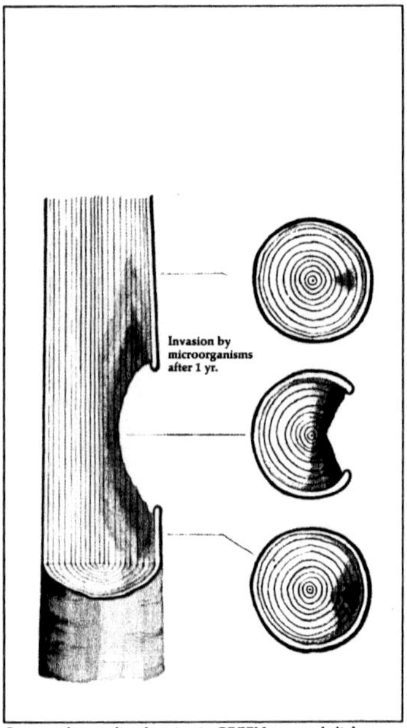

One year later—discoloration in GREEN area and slight discoloration in RED area.

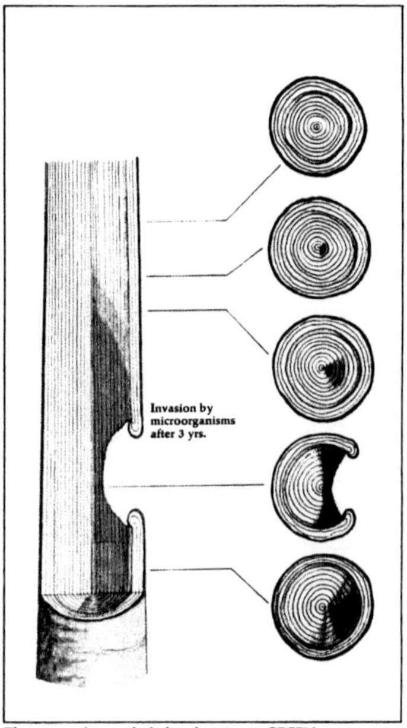

Three years later—dark discoloration in GREEN area; slight discoloration in RED area.

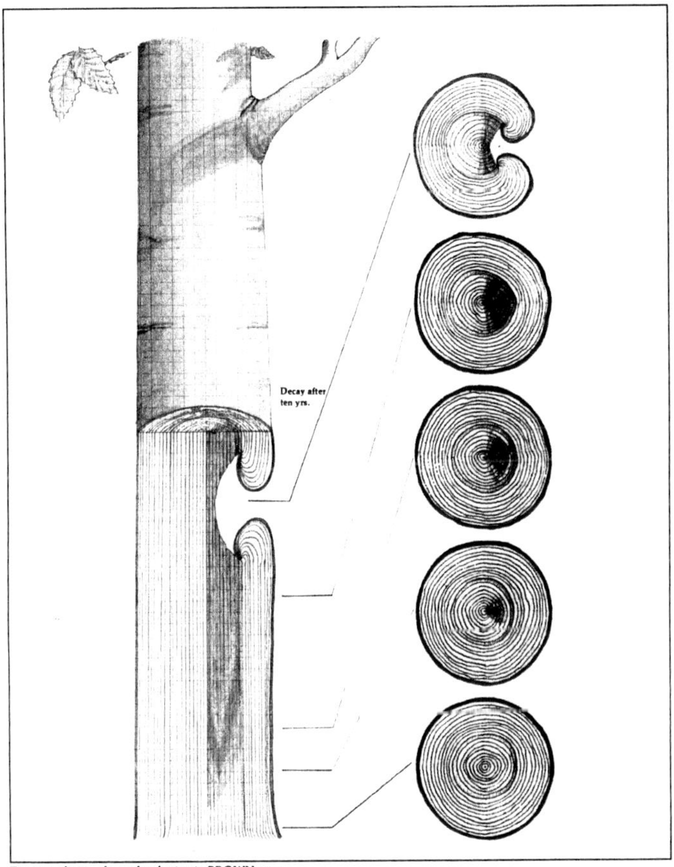

Ten years later—decay developing in BROWN area.
Discoloration in GREEN and RED areas. Decay mostly by
decay fungi; bacteria and nondecay fungi in GREEN area.

The rate of wood decomposition depends on the interactions between the microorganisms and the tree. When the microorganisms are not able to grow through the chemical barriers, the wounds heal. When the microorganisms are able to invade, the tree forms a second line of defense by walling off the invaded tissues. The tree COMPARTMENTALIZES the invaded tissues.

Wound healing is the rule in nature.

Compartmentalization of healed branches.

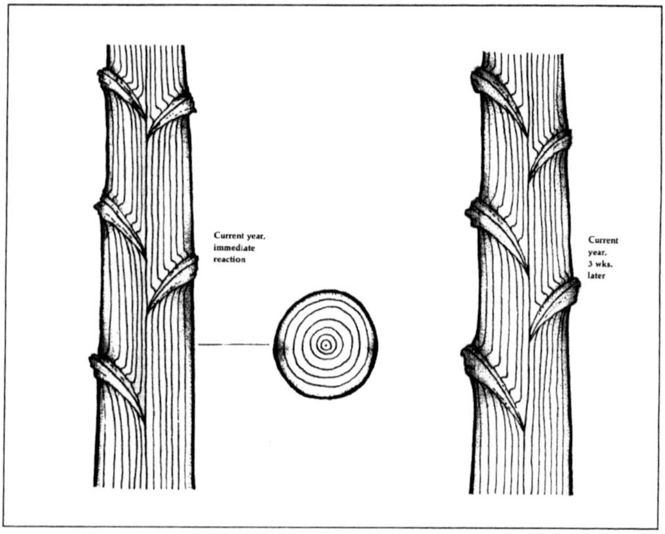

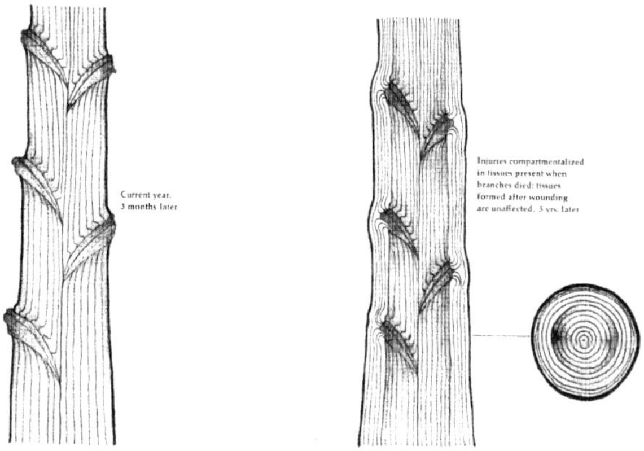
Current year,
3 months later
Injuries compartmentalized
in tissues present when
branches died: tissues
formed after wounding
are unaffected. 3 yrs. later

Compartmentalization of a healed mechanical wound.

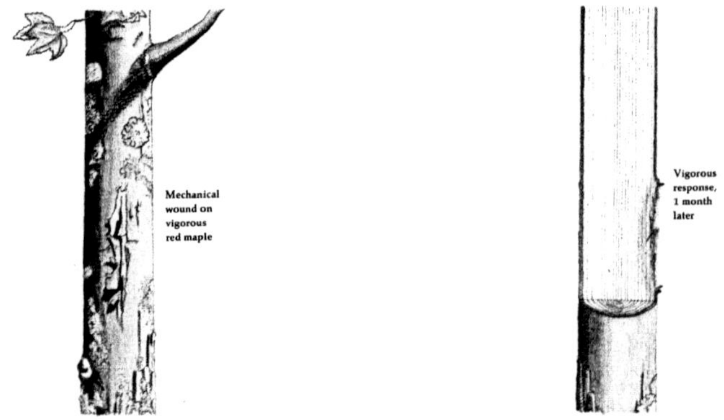

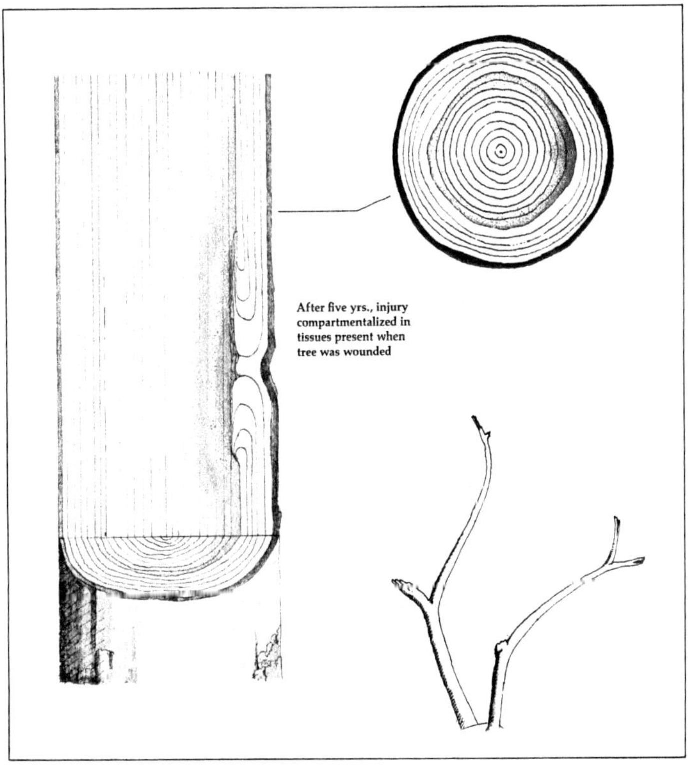
After five yrs., injury
compartmentalized in
tissues present when
tree was wounded

Compartmentalization of poorly healed branches.

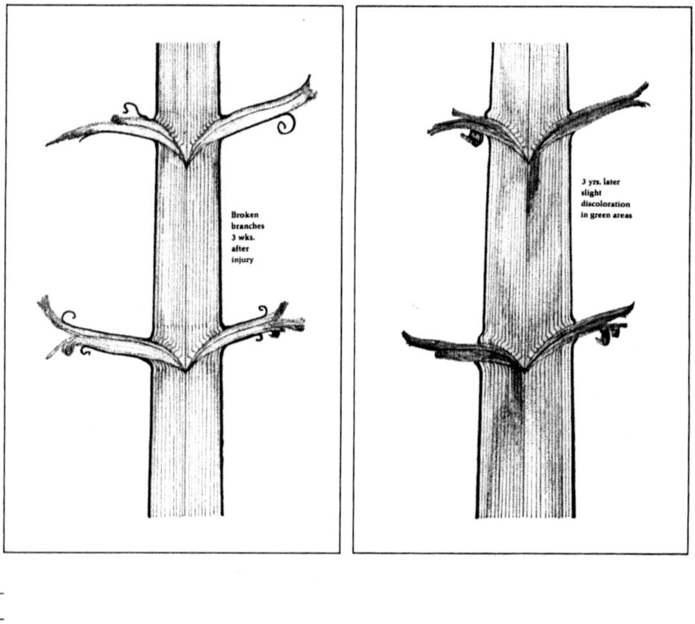

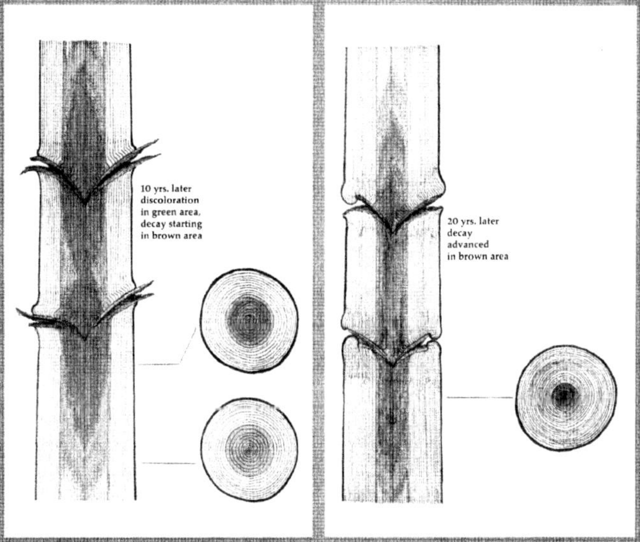
10 yrs. later
discoloration
in green area,
decay starting
in brown area
20 yrs. later
decay
advanced
in brown area

Compartmentalization of poorly healed severe mechanical wounds.

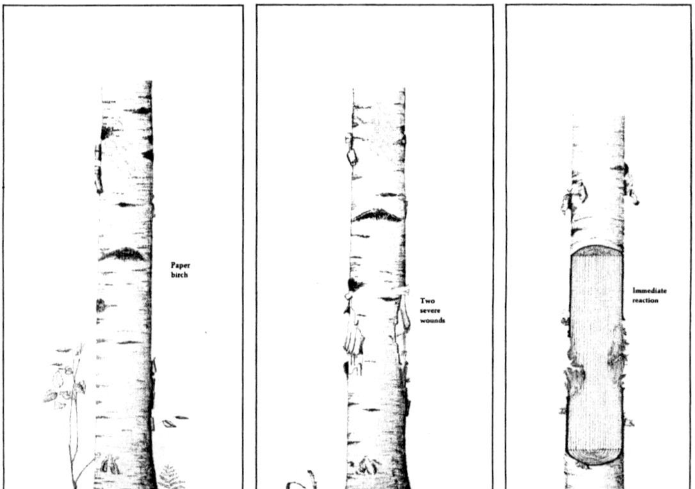

10 yrs. later
discoloration
advanced
decay beginning
20 yrs. later
decay
advanced

What can be done to prevent and minimize damage caused by decay?

HELP THE TREE HELP ITSELF AFTER WOUNDING:

1. **Prevent wounds; use care when working around trees.**
2. **Clean wounds; cut torn bark, and shape wound like an ellipse when possible.**
3. **Sanitation; remove dead, dying, and weak branches from the injured tree.**
4. **Water and fertilize properly.**
5. **Contact professionals for additional advice.**

September 1976

☆U.S. GOVERNMENT PRINTING OFFICE: 1976-0-216-174

For sale by the Superintendent of Documents, U.S. Government Printing Office, Washington, D.C. 20402 - Price $1.30
Stock Number 001-001-00338-7 Catalog Number A13.42/2:T71/2

Woodsy says:

[illegible] use—not abuse.

NE-INF-16-73

Spotted something that isn't quite right?

Let us know and we'll fix it for future reprints. And as a thank you for your diligence, we'll send you a free Sothis book of your choice.

publisher@sothis.press

This page intentionally left blank

Except of course for that line above. And this one. And whilst we have your attention, perhaps you'd like to follow us on Instagram @sothis.press. Go on.

Made in United States
Cleveland, OH
05 November 2026

47673685R00023